Cities
Bangkok

ABDO Publishing Company

Nancy Furstinger

visit us at
www.abdopub.com

Published by ABDO Publishing Company, 4940 Viking Drive, Edina, Minnesota 55435. Copyright © 2005 by Abdo Consulting Group, Inc. International copyrights reserved in all countries. No part of this book may be reproduced in any form without written permission from the publisher. The Checkerboard Library™ is a trademark and logo of ABDO Publishing Company.

Printed in the United States.

Cover Photo: Corbis
Interior Photos: Corbis pp. 1, 5, 6-7, 8, 11, 12, 13, 14, 15, 17, 18, 19, 20, 21, 22, 23, 24, 25, 26, 27, 29

Series Coordinator: Jennifer R. Krueger
Editors: Jennifer R. Krueger, Megan Murphy
Art Direction & Maps: Neil Klinepier

Library of Congress Cataloging-in-Publication Data

Furstinger, Nancy.
    Bangkok / Nancy Furstinger.
       p. cm. -- (Cities)
    Includes index.
    ISBN 1-59197-852-1
      1. Bangkok (Thailand)--Juvenile literature. I. Title.

DS589.B2F87 2005
959.3--dc22

                                                  2004053715

CONTENTS

Bangkok 4
Bangkok at a Glance 6
Timeline 7
Early Bangkok 8
City of Kings 10
Government 12
River of Kings 14
World Port 16
Monsoons 18
Thai Culture 20
Temples 24
Leisure 26
Sights to See 28
Glossary 30
Saying It 31
Web Sites 31
Index 32

BANGKOK

For a capital city, Bangkok is young. It has only been the capital of Thailand for about 200 years. In that time, Bangkok changed from a village of wooden huts into a modern city.

Bangkok lies on Thailand's chief river, the Chao Phraya. About 25 miles (40 km) downstream from Bangkok, the river empties into the Gulf of Thailand. Bangkok is only six feet (2 m) above sea level. Floods often threaten the low-lying parts of the city during the rainy season.

Canals form a royal island in the heart of the city. This is the old city center. The royal palace and many older temples are located there. Nine kings have ruled over Bangkok from the island palace.

This Thai kingdom is a golden wonderland. Hundreds of gleaming temples display statues of **jade** and gold. Monks emerge single file from the temples at dawn. During a water festival, thousands of tiny, candlelit boats float past the Grand Palace.

Opposite Page: *More than 200 years ago, Bangkok was known as the Village of the Wild Plum. Today, this capital city is home to more than 7 million people.*

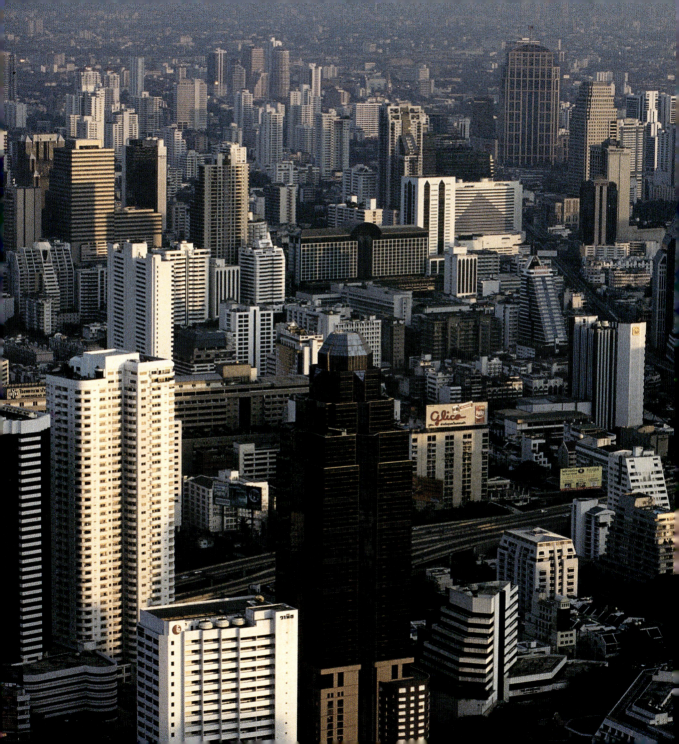

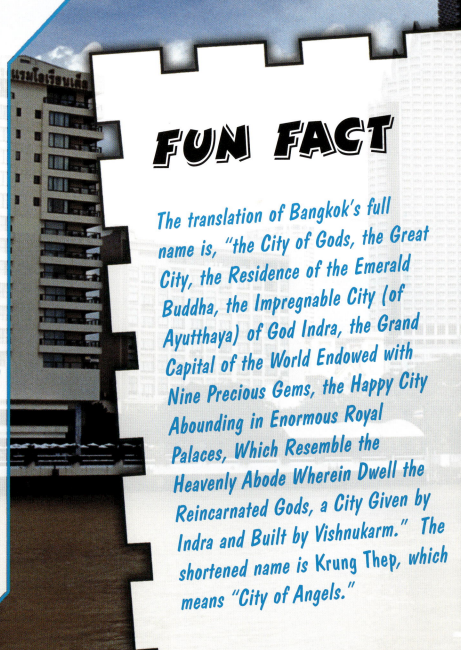

BANGKOK AT A GLANCE

Date of Founding: 1782
Population: 7 million
Metro Area: 612 square miles (1,585 sq km)
Average Temperatures:
- 77° Fahrenheit (25°C) in cold season
- 86° Fahrenheit (30°C) in warm season

Annual Rainfall: 60 inches (152 cm)
Elevation: 0–9 feet (0–3 m)
Landmarks: Grand Palace, wats
Money: Baht
Language: Thai

FUN FACT

The translation of Bangkok's full name is, "the City of Gods, the Great City, the Residence of the Emerald Buddha, the Impregnable City (of Ayutthaya) of God Indra, the Grand Capital of the World Endowed with Nine Precious Gems, the Happy City Abounding in Enormous Royal Palaces, Which Resemble the Heavenly Abode Wherein Dwell the Reincarnated Gods, a City Given by Indra and Built by Vishnukarm." The shortened name is Krung Thep, which means "City of Angels."

TIMELINE

1782 - Bangkok is named the capital of Siam; Rama I rules as the first king of the Chakkri dynasty.

1851–1868 - The most famous Thai king, Rama IV, reigns.

1917 - Chulalongkorn, Thailand's oldest university, is founded.

1938 - Rama VIII grants part of Dusit Palace's land for use as a public zoo.

1939 - The country's name changes from Siam to Thailand.

1946 - Rama IX takes the throne of Thailand.

1997 - Thailand adopts a new constitution.

EARLY BANGKOK

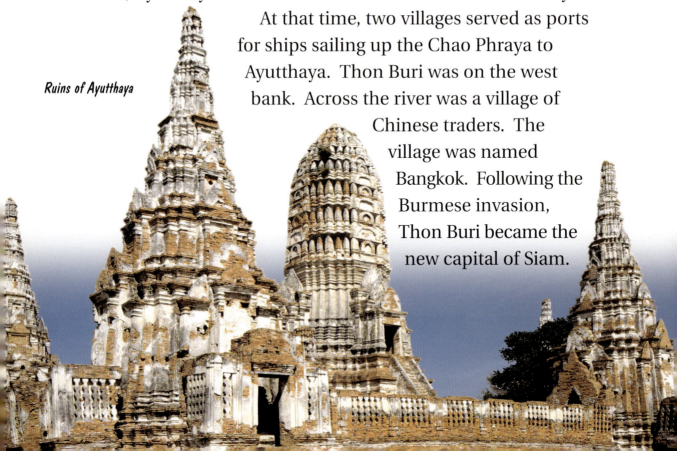
Ruins of Ayutthaya

In the 1700s, Thailand was called Siam. Its capital was Ayutthaya. Siam was often at war with the neighboring kingdom of Burma. Burma is now known as Myanmar. In 1767, Ayutthaya fell under the control of the Burmese army. At that time, two villages served as ports for ships sailing up the Chao Phraya to Ayutthaya. Thon Buri was on the west bank. Across the river was a village of Chinese traders. The village was named Bangkok. Following the Burmese invasion, Thon Buri became the new capital of Siam.

The Siamese eventually recovered control from Burma. In 1782, the capital of Siam changed again. The new king, Rama I, was worried that the Burmese army would attack again. The king wanted to block his enemy's path with water. So, he moved the capital across the river to Bangkok.

But Rama I didn't stop there. First, he built his Grand Palace at a bend in the river. Then the king had several *khlongs*, or canals, dug. These canals created an island around the palace that could be easily defended. The *khlongs* also allowed boats to navigate around Bangkok.

CITY OF KINGS

Rama I was the first ruler of the Chakkri **dynasty**. He took the throne in 1782. Perhaps the most famous king was Rama IV, better known as King Mongkut. During his rule from 1851 to 1868, he modernized Siam. And, he introduced Western **culture** to Bangkok.

In 1939, Siam's name was changed to Thailand. Before that time, the country had never been ruled by a foreign power or nation. The Thai call their country *Muang Thai*, or "Land of the Free." They are very loyal to their kings.

Thailand's king serves as chief of state. He appoints a **prime minister** to serve as the head of government. Today's king, Bhumibol Adulyadej, was born in Massachusetts. He took the throne as Rama IX in 1946. He is the longest ruling monarch in Thai history.

Although Thailand has a king, the country is also served by a **democratic** government. It has been a democracy for more than 70 years. In 1997, a new **constitution** was adopted by the Thai government. The new document gives more political power to the people.

BANGKOK'S KINGS

The Chakkri dynasty is Thailand's ruling house. It was founded by Rama I, Chao Phraya Chakkri, in 1782. His descendants reigned in an unbroken line after him.

Rama II - 1809–1824
Rama III - 1824–1851
Rama IV - 1851–1868
Rama V - 1868–1910
Rama VI - 1910–1925
Rama VII - 1925–1935
Rama VIII - 1935–1946
Rama IX - 1946–

Rama IX and his only son, Crown Prince Vajiralongkorn, in 1978. The prince is being ordained as a monk.

GOVERNMENT

Thailand is divided into 76 **provinces**. A governor controls each province. In Bangkok, the governor is elected by popular vote for a four-year term. The governor appoints four deputy governors to help perform his or her duties.

Bangkok's government also includes a council of elected members. The number of members is based on the city's population. One councillor represents 100,000 people. The council makes laws for the city. It also manages the budget.

Bangkok is the only province that elects its governor. The other 75 provinces have their governors appointed by the country's Ministry of the Interior. Here, a woman drops her ballot in the box during an election for Bangkok's governor.

The Royal Guard performs in front of the Parliament Building in Bangkok.

As a capital city, Bangkok is home to Thailand's government as well. The palaces and mansions surrounding the Grand Palace used to house the government offices. However, space became short. The country's officials were forced to move to large buildings and offices nearby.

RIVER OF KINGS

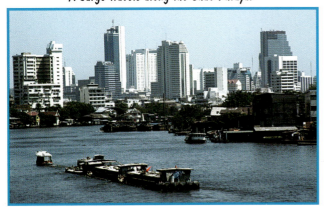

A barge travels along the Chao Phraya.

The Chao Phraya, or River of Kings, forms a watery highway to the west of Bangkok. In the old city, the canals served as streets and avenues. People often traveled more by water than by land.

As Bangkok grew, many of these canals were filled in to form streets. Now buses, three-wheelers called *tuk-tuks*, cars, and motorcycles create the most traffic. In the 1970s, city authorities established a bus system and opened a sky train. However, heavy traffic remains a major problem in Bangkok.

Because of Bangkok's location on the river, people still travel by water. Boats of all kinds float through its waterways. Water taxis zip past barges pulling mounds of rice. Riverboats link Bangkok with all the main towns along the river.

A water taxi cruises along the Chao Phraya. The Temple of Dawn, or Wat Arun, stands on the southern bank of this busy river.

WORLD PORT

Bangkok is Thailand's center of industry and trade. About one-third of the goods produced in Thailand come from Bangkok. Nearly half of the nation's businesses are located there. Most of its banks are located in the capital as well.

Bangkok is surrounded by the Chao Phraya **delta**. The delta creates a large, flat, and extremely fertile area. More rice is grown here than in most places in Asia. So, it is sometimes called "the rice bowl of Asia."

The natural and man-made canals within the delta all connect to the Chao Phraya. The river then flows into the Gulf of Thailand. These waterways have made Bangkok Thailand's main international port.

The city exports rice, tapioca, corn, sugarcane, rubber, and fish. People process foods such as rice and shrimp. They make chemicals, fabrics, and computer parts. Teak was once an important resource, too. But, logging has been banned due to the shortage of this native tree.

Bangkok's river and canals carry the famous floating markets. Each morning, hundreds of boats crowd the waterways. Vendors sell fruits, vegetables, fish, and flowers.

Tourism is also a major source of income in Thailand's capital. Bangkok's markets showcase many of the **unique** treasures the city has to offer. Stalls sell fresh crabs, chilies, vases, and fragrant flowers. Tourists seek woven silks, potteries, and statues. The city is also famous for its jewelry trade.

MONSOONS

In addition to its jewelry, Bangkok is also famous for its beautiful silks. Bangkok's climate is perfect for this industry.

A basket of silkworms

The mulberry tree flourishes in Thailand's weather. The leaves of this tree feed silkworms. These worms spin a silken cocoon. Thai people spin and dye the silk strands. Then, they weave these threads into beautiful cloth.

From May to October, drenching downpours mark the rainy season in Bangkok. **Monsoons** blow from the southwest, bringing most of the annual rainfall. The winds switch direction in November. The northeast monsoon brings a cool, dry season.

The average annual rainfall is about 60 inches (152 cm). The **delta** surrounding Bangkok is low-lying, so the city sometimes floods during the rainy season. However, the hot season often brings long periods of dry weather.

A Thai woman weaving silk

THAI SILKS

An American architect and intelligence officer named Jim Thompson became interested in Thai silk weaving. Thompson brought samples of the vibrantly colored silk back to New York in 1947 and began selling them. He founded the Thai Silk Company in 1948. Soon, his company became a worldwide business.

The treasures Thompson found can still be viewed at the Jim Thompson House in Bangkok. However, the man who made silks famous is nowhere to be found. He vanished while on vacation in Malaysia in 1967. His whereabouts remain a mystery.

THAI CULTURE

The people of Bangkok are known for their friendliness. In fact, Thailand is often called the "land of smiles." Thai usually bow with their hands together as a formal greeting. In a city of 7 million people, being polite is important.

Most houses in Bangkok are small, one- or two-story wooden houses. Many Thai live in apartments above shops. Overcrowding and housing shortages are common problems.

Thai is the national language, but Bangkok schools also teach English as a second language. Chinese is widely spoken among the Chinese **immigrants**.

People enjoy Thai food all over the world. It is often seasoned with garlic and chilies. Citrus and fish sauces also give Thai food its flavor. Thai eat rice with most meals. Noodle and **curry** dishes are

Traditional Thai dishes

Chinese influence can be seen throughout Bangkok. Here, a Chinese temple pavilion sits along the Chao Phraya.

CITY OF TRADERS

Before the start of the Chakkri dynasty, Bangkok was a village of Chinese traders. Then, King Rama I moved his capital from Thon Buri to Bangkok in 1782. At that time, the Chinese merchants were forced to leave the area intended for the Grand Palace. They moved to the present location of Chinatown.

In the 1800s, more Chinese immigrants arrived in Bangkok looking for work. By the early 1900s, they owned most of the businesses in Bangkok. After World War II, the number of Chinese immigrants decreased because of stricter immigration laws. Today, most of the Chinese who live in Thailand embrace the Thai culture as their own.

also very common. Thai cook with vegetables, chicken, pork, seafood, and some beef.

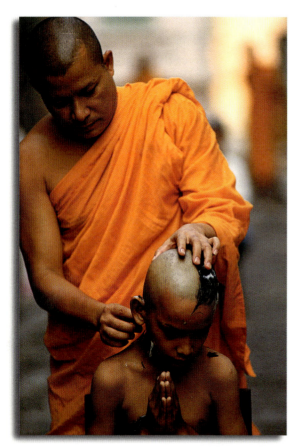

A monk shaves the head of a novice.

Islam, Christianity, and Hinduism are all practiced in Bangkok. However, most people honor **Buddhist** traditions. Most young Thai males become Buddhist monks. They shave their heads and eyebrows and wear orange robes. The time they spend training ranges from a few days to a few months.

The Thai government oversees education in Bangkok. There are four levels of formal education, including preschool, primary, lower and upper secondary, and higher education. Despite overcrowding in schools, **literacy** is high.

The Thai government also funds the college education of students in Bangkok. The oldest university in Thailand, Chulalongkorn, was founded in 1917. Students study medicine at Mahidol. The Temple of the Great Relic contains one of Thailand's two **Buddhist** universities.

Bangkok schoolchildren

TEMPLES

Outside Wat Suthat stands the 80-foot (24-m) Giant Swing. This teak swing honors the Hindu god Shiva. Swinging contests used to be held here. A participant would attempt to grab a bag of money with his or her teeth. These perilous contests were outlawed in the 1930s.

There are more than 300 **Buddhist** temples in Bangkok. These buildings highlight why Thailand's capital is called the "Divine City." Typical temples, or *wats*, contain monks' living spaces, meeting centers, and shrines. These shrines honor Siddhartha Gautama, or Buddha, the founder of Buddhism.

At Wat Phra Kaeo, the Emerald Buddha sits high on a pedestal. Carved from **jade**, it is Thailand's most valued statue of Buddha. Tourists are not allowed to take pictures of it. Wat Phra Kaeo is connected to the Grand Palace and was built the same year Bangkok was founded.

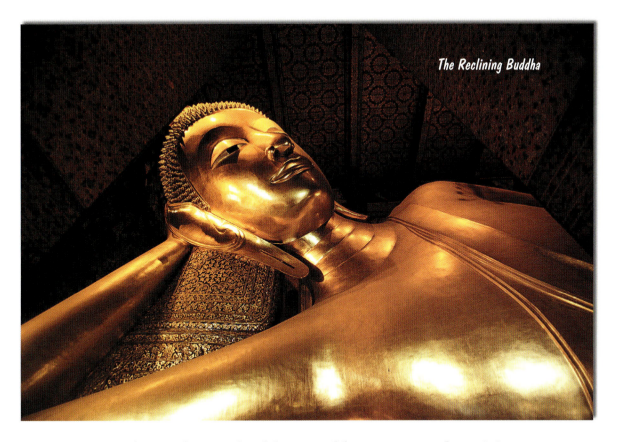

The Reclining Buddha

Wat Pho is the city's oldest and largest temple. Visitors can learn **meditation** there. It is also a center for Thai massage. Wat Pho houses a giant statue called the Reclining Buddha. Covered in gold leaf, he hints at Nirvana, or great bliss.

Wall paintings of Buddha's life are found in Wat Arun. Known as the Temple of Dawn, this *wat* has a tower that rises 269 feet (82 m) in the air. During certain festivals, the outline of the tower is lit up by hundreds of lights.

LEISURE

Loy Krathong, a water festival, takes place during a full moon, usually in late October or November. People fill boats made of banana leaves with incense and candles. Then, they float the boats on ponds, rivers, and canals.

The Thai people love to celebrate. There are many festivals and celebrations throughout the year in Bangkok. One festival honors water spirits and celebrates the end of the rainy season. Another celebrates the start of the rice planting season.

Other events celebrate people. On August 12, the queen's birthday is marked with glittering lights on the Grand Palace.

The king is honored on December 5. Another annual event is the Trooping of the Colours. During this festival, the Royal Guards pledge loyalty to their king on the Royal Plaza.

When they aren't celebrating, many Bangkok residents take part in sports. One popular sport is Thai boxing. It is also known as "the science of eight limbs." This type of boxing started out as a **martial art**. Boxers use elbows, knees, feet, and hands to defend themselves.

Thai kite flying is another popular sporting activity. Flying often takes place near the Grand Palace. Flyers of small kites try to tangle and ground the big fighter pilot kites. People even glue jagged glass to the strings of the small kites to bring the large kites down more quickly.

A Thai boxer prepares for a match with a ceremonial dance.

SIGHTS TO SEE

Perhaps one of the most splendid sites in Bangkok is the Grand Palace. It was once the home of Thailand's kings. Now this gated kingdom is only used for ceremonies. The palace also contains the Royal Thai Decorations and Coin Pavilion. Medals and coins there date back hundreds of years.

Another royal mansion called Vimanmek Palace also houses priceless treasures. This is the largest teak mansion in the world. Inside, a handicraft museum displays silverware, glassware, and ivory.

A favorite tourist attraction is Dusit Zoo. It was the first zoo in Thailand. It was formerly part of Dusit Palace. In 1938, Rama VIII granted the land for use as a public zoo. The Royal Elephant National Museum is also on the grounds of Dusit Palace. It was originally the site of the royal elephant stable.

Tourists explore Thai **culture** at the National Museum. Its collections feature wood carvings, gold treasures, musical instruments, and decorative art. There is even an entire building, the Red House, that was once home to a princess of Rama I.

These attractions make Bangkok one of the most **unique** cities in Southeast Asia. From busy markets to quiet temples, Bangkok is a fascinating and beautiful place. Bangkok has come a long way from its beginnings as a small village of trader huts!

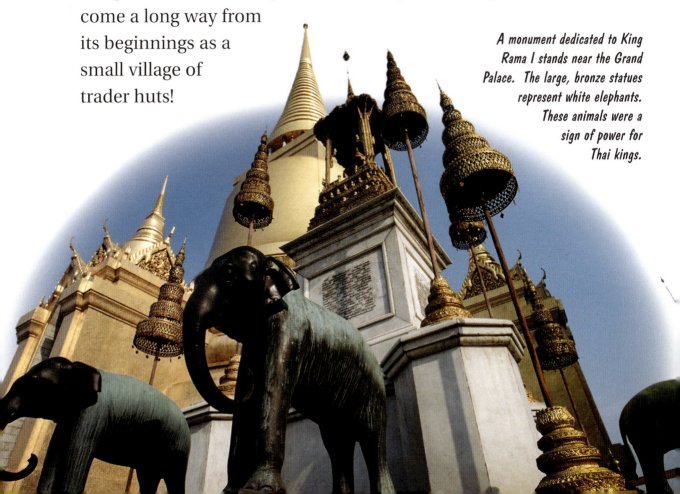

A monument dedicated to King Rama I stands near the Grand Palace. The large, bronze statues represent white elephants. These animals were a sign of power for Thai kings.

GLOSSARY

Buddhism - a religion founded in India by Siddhartha Gautama. It teaches that pain and evil are caused by desire. If people have no desire they will achieve a state of happiness called Nirvana.

constitution - the laws that govern a country.

culture - the customs, arts, and tools of a nation or people at a certain time.

curry - a powdery mixture of strong-smelling spices.

delta - the area of land at the mouth of a river that is formed by the deposit of sediment, sand, and pebbles.

democracy - a governmental system in which the people vote on how to run their country.

dynasty - a series of rulers who belong to the same family.

immigration - entry into another country to live. A person who immigrates is called an immigrant.

jade - a green gemstone.

literacy - the state of being able to read and write.

martial art - a sport practiced for combat or self-defense.

meditation - quiet, careful consideration and thinking.

monsoon - a season of wind that sometimes brings heavy rain.

prime minister - the highest-ranked member of some governments.

province - a geographical or governmental division of a country.

unique - being the only one of its kind.

SAYING IT

Ayutthaya - ah-yoo-TEYE-uh
Bhumibol Adulyadej - POO-mee-pohn ah-DOON-luh-dayt
Chakkri - CHAHK-kree
Chao Phraya - chow PREYE-uh
Chulalongkorn - CHOO-LAH-LAWNG-KAWN
Mahidol - mah-hee-DAWL
Siddhartha Gautama - sihd-DAHR-tuh GOW-tuh-muh
Thon Buri - tun BOOR-ee
Vimanmek - wee-mon-MAKE
Wat Arun - waht a-ROON
Wat Pho - waht PO
Wat Phra Kaeo - waht prah KAY-oe

WEB SITES

To learn more about Bangkok, visit ABDO Publishing Company on the World Wide Web at **www.abdopub.com**. Web sites about Bangkok are featured on our Book Links page. These links are routinely monitored and updated to provide the most current information available.

INDEX

A
attractions 24, 25, 28, 29
Ayutthaya 8

B
Bhumibol Adulyadej 10
Burma 8, 9

C
canals 4, 9, 14, 16
Chao Phraya 4, 8, 9, 14, 16, 18
Chinese 8, 20
climate 4, 18, 26

E
economy 16, 17, 18
education 20, 22, 23

F
festivals 4, 25, 26, 27
flooding 4, 18
food 14, 16, 17, 20, 22

G
government 10, 12, 13, 22, 23
Grand Palace 4, 9, 13, 24, 26, 27, 28

H
housing 20

I
immigrants 20

L
language 20

M
monks 4, 22, 24

R
religion 22, 23, 24, 25
rulers 4, 9, 10, 26, 27, 28, 29

S
Siam 8, 9, 10
Siddhartha Gautama 24
sports 27

T
temples 4, 23, 24, 25, 29
Thailand, Gulf of 4, 16
Thon Buri 8
transportation 4, 8, 9, 14

1754

959.3
FUR

Bangkok.